Bruno Souza dos Santos
Suelen Silva Figueiredo Andrade

Study of the use of stone dust as a fine aggregate in concrete

Bruno Souza dos Santos
Suelen Silva Figueiredo Andrade

Study of the use of stone dust as a fine aggregate in concrete

Evaluation of the compressive strength and water absorption of concrete with stone dust as fine aggregate

ScienciaScripts

Imprint
Any brand names and product names mentioned in this book are subject to trademark, brand or patent protection and are trademarks or registered trademarks of their respective holders. The use of brand names, product names, common names, trade names, product descriptions etc. even without a particular marking in this work is in no way to be construed to mean that such names may be regarded as unrestricted in respect of trademark and brand protection legislation and could thus be used by anyone.

Cover image: www.ingimage.com

This book is a translation from the original published under ISBN 978-620-6-76012-2.

Publisher:
Sciencia Scripts
is a trademark of
Dodo Books Indian Ocean Ltd. and OmniScriptum S.R.L publishing group

120 High Road, East Finchley, London, N2 9ED, United Kingdom
Str. Armeneasca 28/1, office 1, Chisinau MD-2012, Republic of Moldova, Europe
Printed at: see last page
ISBN: 978-620-8-07830-0

Contents

DEDICATORY

I dedicate all this work to God, because he is the sovereign Lord of all my existence, I live to serve him and worship him, I would have nothing without him; to my parents for being the structure that sustains me and responsible for guiding me and teaching me to walk the right path in life.

ACKNOWLEDGEMENTS

To God for all my achievements so far and for allowing me to get this far, planning and looking after my life and my family with an inexplicable love and care. For getting me through all the difficulties and showing me day after day that no matter what happens, He will always be present in our lives.

To my family, my father Jaiuson, my mother Nina, my sister Bruna and my girlfriend Lavínia who have always supported me and spared no effort from an early age so that I could have access to a quality education, for being my base and the origin of all my dreams, without forgetting the unconditional love that is present in our relationship as a family. I am grateful to God for my family.

To my grandparents, uncles, aunts, cousins, who all contributed in their own way, whether near or far, positively influencing this journey, thank you all so much!

To my supervisor Professor Drª Suelen Silva Figueiredo for all the support she has given me, the patience she has shown in all the conversations and the learning she has passed on. For having been present and willing to help from the beginning of the work, for the incentives and all the support at the beginning, middle and especially at the end of my journey at this institution, undoubtedly a partnership that went beyond the master-student barrier, becoming a great friend that I will carry in my heart for the rest of my life.

To the collaborators of the research project, Raquel and Danylo, who were of the utmost importance in the realisation and completion of this work.

To all my friends I've made along the way: Herculys, Ricardo, Myllena, Dona Malba, Emanoel, João Paulo, Anderson, Denns, Adriano, Carlos, Marcos André, my roommate Mattheus Henrique and all my colleagues who have encouraged me, supported me and given me the strength to continue the journey, and shared with me the most varied feelings that only academic life can provide.

I would like to thank all the teachers, technicians and staff who helped with this project, especially Emanoel, the lab manager, who was there for me on several occasions during the time I spent carrying out various laboratory tests.

UFCG for making it possible to carry out and develop the research.

To everyone who directly or indirectly contributed to the completion of another stage in my life.

SUMMARY

The high consumption of natural resources by the construction industry has led to increased concerns about environmental preservation, thus motivating numerous studies on the use of waste that can fully or partially replace certain materials, including aggregates, thereby reducing the environmental impact caused by the waste generated by the construction industry. Among the materials most commonly used to replace fine aggregate is stone dust, a waste product from the rock crushing process. Its use is very promising, given the high cost of natural sands, their scarcity near major centres, as well as the environmental damage their exploitation causes to the environment. In this context, with a view to removing waste from nature and applying it to alternative products, this research project sought to analyse the feasibility of using stone dust as a total and partial substitute for concrete aggregate. The physical characterisation of the raw materials used was carried out: particle size, unit mass, specific mass and fineness. After the physical characterisation, the mix and dosage of the concrete were established. The concrete was moulded into cylindrical specimens with a diameter of 100 mm and a height of 200 mm, and placed in submerged curing at ages of 7, 14 and 28 days. The proportions of waste used were 25%, 50%, 75% and 100% as a substitute for sand, and specimens were also moulded with conventional concrete to serve as a comparative parameter. Finally, the mechanical tests of simple compressive strength and water absorption were carried out, verifying that for all proportions of incorporated stone dust, the concrete showed satisfactory compressive strength results, despite having higher water absorption, with the best strength being associated with the proportion of 100% replacement while also obtaining higher water absorption.

Keywords: Waste; Concrete; Construction; Environmental Impact.

1 INTRODUCTION

Before the great development and growth of the world's population, as well as the increase in natural resources and energy, human beings in their early days survived on the basis of tools and elementary activities, which for the time, even though their waste was released into the environment, did not generate major impacts. With technical advances and the emergence of industrialisation (large-scale production of consumer goods and the introduction of new packaging on the market), the proportion of waste grew and then began to be of various kinds, a fact that influenced the ongoing process of environmental deterioration, causing serious implications for human quality of life (THEISEN, 2012).

From then on, the environmental problem involving the generation of waste took on proportions of vast magnitude, so discussions aimed at reducing the production of this waste, as well as ways of disposing of it, became a priority.

Today, the construction industry is among the human activities responsible for generating a great deal of waste. At the same time, it represents one of the sectors of the economy that continues to develop despite the global recession. According to data from the IBGE's Annual Survey of the Construction Industry, in 2016 the sector was responsible for generating R$319.6 billion in added value (VA) in Brazil, this figure representing 29.5 per cent of the total VA produced in the national economy, remembering that added value is an approximation of the sector's GDP, a sort of GDP of the construction industry net of indirect taxes.

In the same year, the IBGE states that the Northeast was the region that most recorded a gain in participation from 2007 to 2016 in the share of VA that the construction industry has in the national economy, from 19.5 per cent in 2007 to 21.4 per cent in 2016. Despite this, according to Dias (2010), it can be seen that this sector, amid the growth in environmental impacts caused by human activities, is the sector in which the generation and accumulation of huge quantities of waste is prevalent and growing, and which, as a rule, is indiscriminately discarded in the environment.

The construction industry is taking advantage of the good results it has accumulated in recent years, which have acted as a driving force in the face of the impacts they are having as a result of the downturn in the national economy. According to Agopyan (2013), the construction industry is the most polluting industry and is responsible for consuming between 40% and 75% of raw materials 7 produced on the planet. Currently, cement consumption is greater than food consumption and concrete consumption is second only to water consumption. For every human being, 500 kilos of rubble are produced, which is equivalent to 3.5 million tonnes a year.

Raw materials classified as non-renewable mineral goods, such as fine and coarse aggregates, are used to make concrete, which is commonly made up of a mixture of cement, natural sand, gravel and water. In view of this, Carlos (2018) proposed the use of mineral waste in industrial sectors that can incorporate concrete or mortar, which will at least partially replace aggregates in the production process, provided that quality is maintained and/or new products with equivalent properties or differentiated and environmentally friendly products are produced.

Waste has a low cost compared to natural aggregate, which is increasing as the extraction of natural sand is being strictly controlled and frequently curbed by environmental agencies, given the great increase in its scarcity. In addition to reducing the final cost of concrete, this is beneficial for the environment.

Among the waste products, one stands out as the main alternative for replacing natural fine aggregates: stone dust from rock crushing, which can be incorporated into both concrete and mortar, partly replacing fine aggregates, can be an important and technically viable alternative.

Andriolo (2005) defines stone dust as fine material, with particles less than 4.8 mm in diameter, obtained by crushing rock. This crushed sand has a very homogeneous particle size distribution due to the comminution process adopted. However, the grains, like gravel, have an angular shape and rough surface that can reduce workability in concrete. When they are cubic in shape, such as those from granite, stone and with a high proportion of silica, they allow for greater interaction with the mix, resulting in a reduction in the amount of voids in the concrete.

The main function of the fine aggregate in concrete is as a filler, i.e. it is an inert material, so replacing it with materials with similar characteristics does not compromise the final quality. However, the size and geometry of the grains define important characteristics of the material, such as resistance and workability, and must therefore be taken into account in the studies carried out.

The aim is to study the influence of the total and partial replacement of fine aggregate in concrete with stone dust, in order to compare the performance of the concrete on different fronts, such as compressive strength, water absorption and workability, in order to suggest a different destination for the waste, reducing the environmental impacts caused by its improper disposal.

2 OBJECTIVES

2.1. General Objective

To study the technical feasibility of replacing natural fine aggregate with stone dust in the production of concrete.

2.2. Specific objectives

- Characterise conventional and alternative raw materials in physical terms.
- Technologically characterise the physical-mechanical properties of the concrete incorporated with the waste by means of simple compressive strength and water absorption tests.
- Compare the results obtained from concrete incorporated with waste with each other, with conventional concrete and with the values established in the standards.

3 JUSTIFICATION AND RELEVANCE

The technological revolution and high population growth have resulted in enormous consumption of natural resources and the production of large quantities of polluting waste. Waste production has grown sharply as a result of the manufacture of equipment and the high demand for a better standard of living and comfort sought by the population. In this way, the environment has suffered a great deal of damage caused by human exploitation, which has led to major and tenuous discussions between civil and environmental professionals, aimed at reconciling development with the preservation of the environment.

According to Souza (2007), among the materials most consumed in construction is concrete, which consists of a mixture of cement, sand, gravel and water, resulting in a resistant material that is consumed a lot every day. One of the natural materials most commonly used in the manufacture of concrete is sand, which is now consumed in huge quantities. While this sand is taken almost entirely from lowlands and riverbeds, it has numerous environmental consequences, leading to an increase in its scarcity. Stone dust, which does not have a defined final destination and is generally stored in the open in quarries, thus leading to environmental damage such as atmospheric pollution, silting up of rivers and contamination by leached material in drainage areas, emerges as a possible alternative to be studied as a solution to this social, economic and environmental problem.

Thus, the development of new techniques that incorporate waste into the manufacture of materials brings greater profitability to the sector, as well as making it possible to combat the improper disposal of waste in nature and consequent environmental pollution, as well as reducing the use of natural resources.

Therefore, the main motivation stems from the interest in producing a material that contributes to sustainable construction, reconciling the progress of the construction industry with the preservation of the environment, as well as allowing the quality of life of individuals in the social and economic aspect.

4 THEORETICAL FRAMEWORK

4.1. Concrete

According to Mehta (2008), concrete is a composite material that has an agglomerating paste to which aggregate particles or fragments are agglutinated. According to the American Society for Testing and Materials (ASTM, 1997), concrete is a material consisting of an agglomerating medium in which particles of different natures are agglutinated: the binder is cement in the presence of water; the aggregate is any granular material such as sand, pebbles, crushed rock, blast furnace slag and construction and demolition waste. If the aggregate particles are larger than 4.75mm, it is called coarse; otherwise, the aggregate is called fine. To concrete in its fresh state, we can also add chemical substances, additives, which alter certain properties, adapting the concrete to construction needs.

It can also be defined as "a heterogeneous mixture of cement, fine and coarse aggregates, with or without the incorporation of minority components, which develops its properties through the hardening of the cement paste", as defined by BATTAGIN (2009).

The story of how concrete came about is not a simple one, as it is the result of a complex and controversial process of evolution. Some authors date the first cements and concretes to millions of years ago, from the natural formation of sedimentary rocks. According to Kaiffer (1998), archaeological excavations have revealed traces of a building dating from around 4000 BC that was partly made of concrete.

There are various types of classifications for concrete. According to the ABNT NBR 12655/15 standard, concrete can be classified into three categories according to its specific mass in the hardened state:

- Normal concrete: has a specific mass greater than 2000 Kg/m^3 , but does not exceed 2800 Kg/m3. It usually consists of natural sand, crushed stone or rolled pebbles.
- Lightweight concrete: has a specific mass of no less than 800 Kg/m3, but no more than 2000 Kg/m3. Made up of natural or thermally processed aggregates which

have a low density (expanded clay aggregates, steel slag, vermiculite, slate, synthesised sewage waste and others).

- Heavy concrete: specific mass greater than 2800 Kg/m^3 . Made up of high density aggregates such as barite, magnetite, limonite and hematite.

According to Mehta & Monteiro (1997), concrete can also be classified based on compressive strength criteria (Fck at 28 days), divided into the categories shown in Table 1 below:

Table 1: Classification of concrete in relation to its compressive strength.

Type of concrete	Compressive Strength (Mpa)
Low Strength Concrete	< 20
Moderate Strength Concrete	20 < RC < 40
High Strength Concrete	>40

Source: Mehta & Monteiro (2008).

ABNT NBR 12655/15 also contains various other forms of classification and terminology to identify concrete: mass concrete, foam concrete, shotcrete, aerated concrete, prescribed concrete, dosed concrete, etc. According to the Brazilian Association of Concrete Service Companies (ABESC) (2007), commonly used concretes can be classified according to their application on site, since the success of a construction depends on the correct definition of the type of concrete used.

Concrete has evolved a lot in recent times, based on the real and enormous importance it has acquired over time, leading it to be recognised as the most important structural material today. According to Helene (2007), even though it is the most recent of structural building materials, it can be considered one of the most interesting discoveries in the history of the development of humanity and its quality of life, since it has met the needs of security, housing and fortification, hygiene, transport, education, health, leisure, religion and public works.

As Martin (2005) describes, concrete is one of the essential materials of our civilisation. If the components are selected correctly and the dosage is optimised, the properties of the concrete can be profoundly modified, responding to the needs required. It is possible to act on workability, setting times, density, finish and especially durability and mechanical strength, the latter being the property most valued by design and quality control engineers.

4.2. Aggregates

The materials used as aggregates in the manufacture of concrete are found in the earth's crust, originating from igneous or magmatic rocks such as basalt, granite and diabase; sedimentary rocks such as sandstone, claystone, limestone, gypsum and peat; and metamorphic rocks such as gneiss, marble, schist and phyllite. Aggregates used in civil engineering applications are granular materials that do not have a defined shape or volume. These include crushed stone, gravel and natural sands or sands obtained by grinding rock (La Serna & Rezende, *apud* JACQUES, 2013).

According to Neto (2005), among the origins of rocks, the ones that show the best results are igneous, as they are denser and more compacted, e.g. granites and basalts. Metamorphic rocks such as gneiss and quartzite have good potential as aggregates. Sedimentary rocks are the least suitable for use as aggregates due to their high porosity and lower mechanical resistance, and are only used when the requirements are low, such as sandstone and claystones.

Aggregates can be classified according to their origin, whether natural or artificial. One of the most common ways of classifying aggregates is according to their grain size, i.e. their granulometry. According to ABNT standard NBR 7211/09, coarse aggregate is aggregate whose grains pass through a sieve with a mesh opening of 152 mm and are retained on a sieve with a mesh opening of 4.75 mm. Small aggregate is aggregate whose grains pass through a sieve with a mesh opening of 4.75 mm and are retained on a sieve with a mesh opening of 0.075 mm, in accordance with ABNT NBR NM 248/03. These are used in materials such as concrete, which, together with the fine aggregate, fill the mixture. In addition, the arrangement between the grains of the two types of aggregate directly influences the strength of the material.

In Brazil, natural silica sand is widely used as fine aggregate, while crushed stone is more commonly used as coarse aggregate.

According to Bauer (2008), among the existing fine aggregates, the most widely used in construction is sand, whose granulometry and classification are shown in Table 2.

Table 2: Classification of sand according to grain diameter.

Classification	Grain size (mm)
Fine Sand	0,6 a 0,15
Medium sand	2,4 a 0,6
Coarse sand	4,75 a 2,4

Source: Bauer, 2008.

Aggregate can also be natural or artificial. Natural coarse aggregate (gravel) can also be obtained from rivers or dry quarries, but is rarely used today. Artificial coarse aggregate is obtained by crushing rock. ABNT NBR 7211/09 classifies coarse aggregates for concrete into five granulometric ranges, as shown in Table 3. The trade continues to use the nomenclature of the older versions of the same standard. Gravel 1, for example, has grain sizes predominantly larger than 9.5 mm and smaller than 25.0 mm, and is therefore classified as 9.5/25.0.

Table 3: Commercial and ABNT nomenclature for coarse aggregates.

Nomenclature	NBR 7211/09
Brita 0	4,75/12,5
Brita 1	9,5/25,0
Brita 2	19,0/31,5
Brita 3	25,0/50,0
Brita 4	37,5/75,0

Source: ABNT NBR 7211/09

The aggregate used in the composition of concrete is an inert component that fulfils the function of filler and resistant material, made up of particles that must be cemented together by the paste. For this reason, sand, gravel and crushed stone are the main types of aggregate used in concrete production. Aggregates occupy between 60% and 80% of the total volume of concrete, thus providing great cost savings, as well as favouring many concrete properties in the fresh and hardened state.

In recent years, countless studies into the influence of aggregates in concrete have concluded that the role of this component as a filler was too restrictive; on the contrary, it can act on various properties of fresh and hardened concrete, depending on the characteristics of the aggregates. In addition to the size of the aggregate particles, the shape of the grains is also a determining factor in the process of accommodating them, just as it directly affects workability. The less spherical the grains, the greater the empty space between the particles, resulting in a decrease in packing density (LONDERO *et al.*, 2017).

Therefore, the choice of the type of aggregate to be used must meet the requirements and limits established by technical standards. Therefore, when studying the replacement of any material as fine or coarse aggregate in the production of concrete, it is extremely important to take the aforementioned requirements into account.

4.3. Concrete Properties in the Fresh State

According to Souza (2007), the properties of fresh concrete are workability, consistency, cohesion and exudation. For many authors, workability is a property that combines the other properties and is therefore not treated separately.

The workability of concrete is defined by the ASTM - American Society for Testing and Materials as the property that determines the effort required to handle a quantity of fresh concrete with minimal loss of homogeneity (MEHTA & MONTEIRO, 1994). The fundamental characteristic for a concrete to be well compacted is workability, i.e. the suitability of its consistency to the process used to cast and compact it (GIAMUSSO, 1992).

Workability is not an intrinsic property of concrete, as it must be related to the type of construction and the production methods that will be used on site.

The main factors affecting workability are:

a) Internal factors:

- Consistency
- Concrete mix
- Concrete granulometry
- Aggregate grain shape
- Additives

b) External factors

- Type of Mix
- Type of Transport
- Posting Type
- Type of densification
- Dimensions and reinforcement of the part to be made

Depending on the mixing, transport, placing and compacting process adopted on site, the workability of the concrete must be within the required limits so that there is no segregation and good compaction can occur. Manual or mechanised mixing, transport by wheelbarrow or pump, shovelling or chute, manual or vibratory compacting all require different workability. The influence of these factors does not always occur in the same direction, and this takes on particular importance when all of them are considered together (BAUER, 2003).

Workability has a direct influence on whether concrete is easier to use or not, as it makes it easier for the worker to carry out all the tasks required to handle the material. The workability of fresh concrete is a property that is greatly influenced by other properties, and there is no single test to measure it. Consistency is a property that generates plasticity in concrete, which consequently attributes its own characteristics to workability, in other words, concrete that is easier or less easy to handle for a given purpose. The test that measures consistency, called the Slump Test, is the main reference for estimating workability, as we will study later.

4.4. Concrete Properties in the Hardened State

After setting, the concrete hardens, at which point all the properties related to fresh concrete no longer influence the material, but rather the behaviour of the hardened concrete. When studying this state of concrete, the following properties are taken into account: strength, modulus of elasticity, shrinkage, creep, permeability, durability, carbonation and caking (SOUZA, 2007).

According to Mehta & Monteiro (2008), strength is the measure of the stress required to break the material. In the design of structures, typically used in construction, concrete is considered to be the material best suited to resisting compressive loads, which is why compressive strength is generally specified. As concrete strength is a function of the cement hydration process, which is relatively slow, traditionally concrete strength specifications and tests are based on specimens cured under specific temperature and humidity conditions for a period of 28 days.

The ABNT NBR 5739/18 standard regulates the simple compression test of cylindrical concrete specimens (100x200 mm) to check the mechanical strength of concrete.

The deformation modulus (E) is the relationship between the increase in stress and the increase in deformation, i.e. it is the measure of the deformability of concrete, defined by the relationship between stress and relative deformation. It is also related to the measure of the material's rigidity or resistance to deformation. The modulus of deformation is obtained by loading the specimen and observing the deformation corresponding to each load increment (ANDRIOLO, 1984).

Shrinkage is a property of hardened concrete that represents the ability of the concrete to reduce in size. This property is often confused with cracks resulting from shrinkage itself, which appear during the hardening phase when there is rapid drying due to the early evaporation of the concrete's mixing water.

If a concrete specimen is subjected to constant stress for a long time, it will show plastic deformation. The phenomenon of a gradual increase in deformation over time, under constant tension, is called creep (MEHTA & MONTEIRO, 1994). Creep is a phenomenon that occurs in the mix and therefore increases with the addition of paste to the concrete.

Permeability, in turn, is the property of concrete that allows a fluid to pass through its interior. All concrete absorbs a certain amount of water and is permeable within a certain range. Absorption refers to the process by which concrete retains water in its pores and capillary ducts (ANDRIOLO, *apud* SOUZA, 2007). Absorption is measured by drying a sample to constant mass, then immersing it in water and determining the increase in mass expressed as a percentage of the dry mass. While the permeability of concrete is related to its durability, a low permeability concrete is sought so that the access of aggressive agents that could deteriorate the concrete is prevented.

Mehta & Monteiro (1994) define concrete durability as the ability to resist the action of weather, chemical attacks, abrasion or any deterioration process. Durability can be related to the water/cement ratio, with low durability concrete having a high water/cement ratio.

4.5. Stone Dust

Andriolo (2005) defines stone dust as a fine material with particles less than 0.075 mm in diameter, obtained by crushing rock. According to Bauer (1995), stone dust is a finer material than pebbles and its generic but not strict classification is 0 to 4.8 mm. According to Neves *et al.* (2007), when they come from granite rocks, they generally have a cubic shape and rough surfaces.

This crushed sand has a very homogeneous grain size distribution due to the comminution process adopted. However, the grains, like gravel, have an angular shape and rough surface that can reduce workability in concrete. When they are cubic in shape, such as those from granite, stone and with a high proportion of silica, they allow for greater interaction with the mixture, resulting in a reduction in the amount of voids in the concrete. Crushed sand is practically free of organic and clay impurities, as well as the possible problems they cause (COSTA, 2005). They provide greater adherence to the surfaces of application, however, they make the concrete less workable.

According to Cabral (2007), crushed sand has a large amount of powdery material which, within certain limits, makes it more workable by filling the voids in the cement paste and water. On the other hand, when it is present in large quantities, it is detrimental to the quality of the concrete. This occurs when the powdery material forms a film around the particles, making it difficult for the cement paste to adhere to the aggregates and thus increasing the need for water to maintain the same workability of the concrete.

According to Vieiro (2010), important concrete properties are improved with the presence of pulverised materials in its composition, including: compressive strength, workability, shrinkage, permeability and abrasion resistance. The author also notes that in some cases the concrete has reduced cement consumption.

The use of stone dust as a fine aggregate in concrete, both for economic and durability reasons, has been researched and has generated enormous interest, since quarries will be able to commercialise a waste product that did not add value to the process, caused inconvenience in terms of storage and the environment.

4.6. Studies carried out

According to Andriolo (2005), the use of stone dust in Brazil began in the 1980s, based on technical studies carried out at the Itaipu hydroelectric dam, which showed technical and economic advantages. The author states that further studies were carried out by engineers from Construtora Norberto Odebrecht during the construction of the Capanda dam in Angola, resulting in the use of stone dust in RCC (roller compacted concrete) in the aforementioned dam. Today, in São Paulo, stone dust is already being used in the production of normal and high strength concrete. In the metropolitan region of São Paulo, the use of fine aggregate from riverbeds is becoming increasingly complicated, as the sources of natural sand are located at distances of around 120 kilometres, due to the extinction of the nearest natural reserves. According to SILVA *et al.* (2004), concrete companies in the Federal District no longer manufacture concrete with 100% natural river sand due to the high cost of dredging and transporting it, and for environmental reasons related to sustainability.

The incorporation of waste with no economic value into the manufacture of materials has become increasingly viable for the construction industry, and stone dust, the waste in question, has been the subject of several studies, such as the one carried out by Menossi *et al.* (2004), which analysed concrete with different mixtures varying in stone dust content. Different mixes were made for concretes with characteristic Fck strengths of between 15 and 40 MPa. The base mix produced was 1:3: 3 by mass, and the percentages of stone dust used to replace natural sand were as follows: 25%, 50%, 75% and 100% by mass. The results of the characterisation test showed that the waste has all the physical and chemical characteristics required for use as fine aggregate in concrete. Likewise, concrete with added stone dust showed an increase in compressive strength compared to concrete made with natural sand, particularly at 28 days, this increase reached 66%. The authors concluded that replacing sand with stone dust did not compromise the concrete's strength levels or workability.

Bastos (2003) carried out research into replacing natural sand with stone dust. Concrete was produced using the following mixes: 1:2.03:3.02 and 1:3.2:4.2. The percentages of stone dust used to replace natural sand for each mix were as follows: 15%, 30%, 50% and 70% by mass. When 70% was used as a substitute, the water/cement ratio was reduced by an average of 10%. The author observed from the results that the best performance in all the aspects analysed was achieved with the mixture with 70% as a substitute for natural sand. He concluded that the higher the replacement content, the higher the compressive strength. The author also noted that the substitution led to a reduction in water absorption and consequently a reduction in permeability, due to the reduction in the water/cement ratio.

Tiecher *et al.* carried out research comparing different dosing methods for conventional concrete using 100% artificial sand. In this study, it was found that it is difficult to obtain slumps of more than 80 mm for strengths of more than 40 MPa when using stone dust, making it necessary to use more efficient superplasticising additives.

In the context of the studies presented, the economic and environmental variables benefit from the alternative of using stone dust to replace natural sand, guaranteeing progress in the construction industry without neglecting the environmental problem, nor forgetting to provide an improvement in the properties of concrete, strength and water absorption, which was precisely the main point developed in the research.

5 MATERIALS AND METHODS

5.1. Materials

The materials used were

- Fine aggregate: medium/coarse washed sand, commonly used in construction in the region, from the municipality of São Domingos-PB.
- Coarse aggregate: gravel 1, maximum diameter 19 mm, frequently used in construction in the region, from the municipality of Caicó-RN.
- Binder: CP II F 32.
- Mining waste: stone dust waste from the municipality of Caicó-RN.
- Water: from the Pombal municipal water supply.

5.2. Methods

The procedures followed the sequence shown in Figure 1, described below.

Flowchart 1: Activities carried out.

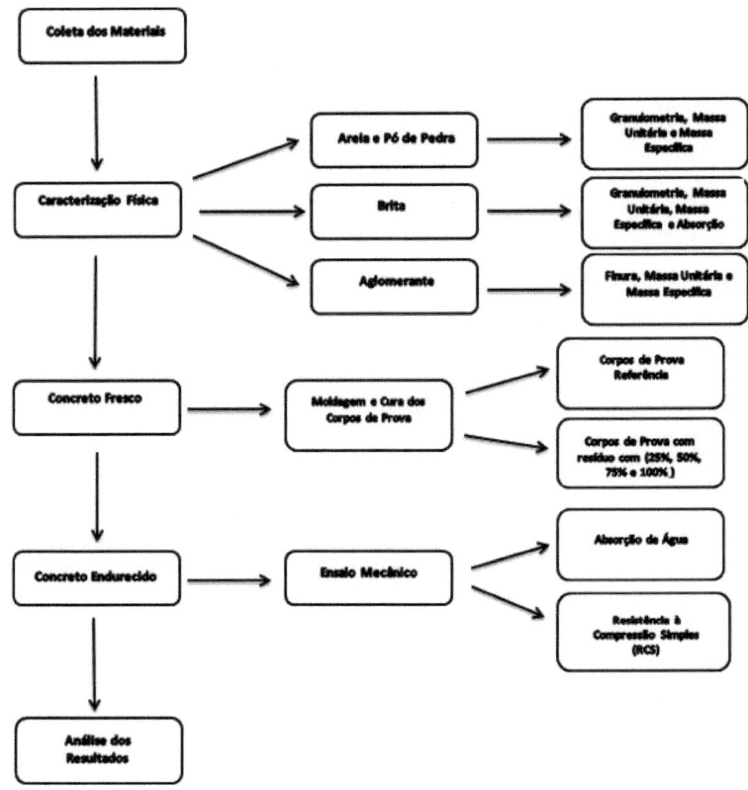

5.2.1. Collection of raw materials

The stone dust waste was collected in the municipality of Caicó-RN, from a crusher in that city. The fine aggregate and coarse aggregate were acquired through the purchase of products commercialised in the municipality of Pombal-PB, coming respectively from the municipality of São Domingos-PB and Caicó-RN. The binder was obtained by buying it from a building materials shop.

Figure 1: Stone dust collection site.

Source: Author, 2019.

5.2.2. Physical Characterisation

The physical characterisation consisted of determining the granulometric composition of the sand and waste, as well as their unit and specific masses. For the binder, in addition to the specific and unit masses, the fineness modulus.

- **Particle size analysis**

Particle size analysis consists of determining the quantity of each grain fraction present in the material, in order to know the dimensions of its particles, since these have a major influence on the outcome of the final product. The procedure was carried out in accordance with ABNT NM 248/03.

Using a mechanical agitator (Figure 3) and a manual process, two 500 g samples of natural aggregate and waste were passed through the sieves of the normal series with openings of 37.5 mm; 19 mm; 4.75 mm; 2.40 mm; 1.20 mm; 0.60 mm; 0.30 mm and 0.15 mm, and the mass of material retained on each sieve was weighed for the two samples and averaged.

Figure 2: Arrangement of the sieves in the mechanical shaker.

Source: Author, 2019.

With the average values of retained mass and accumulated retained mass, it was possible to find the percentage passing through each sieve and its accumulated percentage. Using this data, the aggregate particle size curves were constructed.

- **Unit Mass, Specific Mass and Fineness**

The unit mass of the aggregates was determined according to ABNT NBR NM 45/06. Using a container with a known volume, three samples of the materials were weighed and the average of the values found was subtracted from the mass of the container. The unit mass was obtained by dividing the average mass by the volume of the container.

The specific mass of the sand and waste was obtained according to ABNT NBR NM 52/09 and that of the cement was based on ABNT NBR 16605/17. The fineness of the cement was determined in accordance with ABNT NBR 11579/13.

5.2.3. Concrete Batching and Determination of Design Specifications

Initially, three test mixes derived from the ABCP method were studied: 1:1.97:2.46:0.5, 1:2:3:0.7, 1:3:3.40:0.7, in which specimens measuring 100 mm in diameter and 200 mm in height were moulded with a high content of natural aggregate replaced by stone dust at three curing ages: 7, 14 and 28 days. In conjunction with the moulding, the cone truncation test was carried out on each mix, in accordance with ABNT standard NBR NM 67/98, in order to check the consistency of the concrete. After analysing and investigating the results, it was concluded that regardless of the mixes under study, at high levels of substitution of waste for natural aggregate, the required slump was not obtained, which would cause serious problems in terms of the workability of the concrete.

In view of the situation, studies were carried out and visits made to construction sites in the region that used some waste in concrete as a substitute for natural aggregate, along with research into the literature, and it was decided to use the 1:3:3 mix in mass with a w/c ratio set at 0.7, seeking to investigate the behaviour of the concrete with regard to adequate workability without forgetting the minimum resistance stipulated in the regulations.

5.2.4. Making, moulding and curing the specimens

Once the design had been defined, the test specimens were made. To carry out this stage, the materials were mixed in a concrete mixer, as shown in Figure 4, in the following sequence: first the coarse aggregate, fine aggregate and 50% of the water were added; the mixture was beaten in the mixer for 1 minute; then the cement and the rest of the water were added, beating the mixture for 3 minutes. After 3 minutes, the mixer was turned off so that the slump of the mixture could be measured and the specimens moulded.

Figure 3: Concrete mixer used for moulding.

Source: Author, 2019.

Figure 4: Cylindrical metal moulds ready for use.

Source: Author, 2019.

The specimens were moulded in accordance with the ABNT NBR 5738/16 standard using cylindrical moulds with basic dimensions of 100 mm in diameter and 200 mm in height (Figure 5). During moulding, each specimen was formed from three layers. They were compacted by applying 12 blows to each one, using a compacting rod. The mix used to make these specimens was made with the materials common to concrete (cement + sand + gravel + water) and with the waste that will be used to replace sand, in four proportions: 25%, 50%, 75% and 100%. Conventional specimens were also made, without the addition of waste, in order to serve as a reference and comparison parameter, as well as to verify the properties intended in this work. Submerged curing and open-air curing were used for the specimens, which were subjected to curing ages of 7, 14 and 28 days, as shown in Figures 6 and 7 below.

Figure 5: Tank used for submerged curing of the specimens.

Source: Author, 2019.

Figure 6: Test specimens in open-air curing.

Source: Author, 2019.

Three specimens were moulded for each composition and for each curing age, totalling 135 specimens, of which 45 were for the simple compressive strength (SCR) test with submerged curing, 45 were for the compressive strength (SCR) test with open-air curing and 45 were for the water absorption test.

Figure 7: Moulded specimens.

Source: Author, 2019.

5.2.5. Mechanical tests

After curing the specimens, the mechanical tests for water absorption and simple compressive strength were carried out.

5.2.5.1 Water absorption

Water absorption was determined according to the methodology recommended by ABNT standard NBR 9778/09. The first step of the test consists of drying in an oven at a temperature of 105°C ±5°C until the mass is constant, weighing each dry sample. The saturated masses were obtained after immersion in water at a temperature of 23°C ±2°C for 24 hours. The specimens were then removed from the water, dried and weighed. The percentage difference between the saturated mass and the dry mass corresponds to the total water absorption capacity, calculated on a dry basis.

The individual water absorption values (A), expressed as a percentage (%), were obtained using equation 1, while the average absorption was determined using the arithmetic mean of five repetitions.

$$A = [(M2-M1)/M1]*100 \quad \text{Equation (1)}$$

Where:

M1= mass of oven-dried specimen (g)

M2= mass of saturated specimen (g)

A= water absorption (%)

Figure 8: Test specimens in the oven to obtain the dry mass.

Source: Author, 2019.

5.2.5.2 Simple compressive strength (SCR)

To determine the resistance to simple compression, the specimens were broken in a SHIMADZU AG-IS manual hydraulic press with a 100 KN cell, in accordance with ABNT NBR 5739/18. Figures 10 and 11 show the press used to carry out the test and a broken specimen, respectively.

Figure 9: Hydraulic press used in the axial compressive strength test.

Source: Author, 2019.

Figure 10: Broken specimen.

Source: Author, 2019.

6 RESULTS AND DISCUSSIONS

6.1 . Physical characterisation of materials

6.1.1. Granulometry

Granulometry of the fine aggregate

After collecting and coarsely processing the sand and stone dust, the particle size of the stone dust was analysed, as well as the aggregates, both fine and coarse, in order to identify the size of the grains that make them up. Table 4 and Graph 1 show, respectively, the results of the granulometric composition and the granulometric curve of the stone dust used throughout the research.

Table 4: Particle size composition of stone dust.

ABNT Sieve Opening (mm)	Retained percentage (%)	Cumulative Retained Percentage (%)	Passing percentage (%)
37,5	0,00	0,00	100,00
19	0,00	0,00	100,00
4,75	0,00	0,00	99,60
2,4	13,00	11,00	89,20
1,2	22,00	29,00	71,00
0,6	17,00	43,00	57,20
0,3	19,00	58,00	41,60
0,15	29,00	82,00	17,80
Fund	22,00	100,00	0,00

Source: Author, 2019.

Gráfico 1: Curva Granulométrica do pó de pedra.

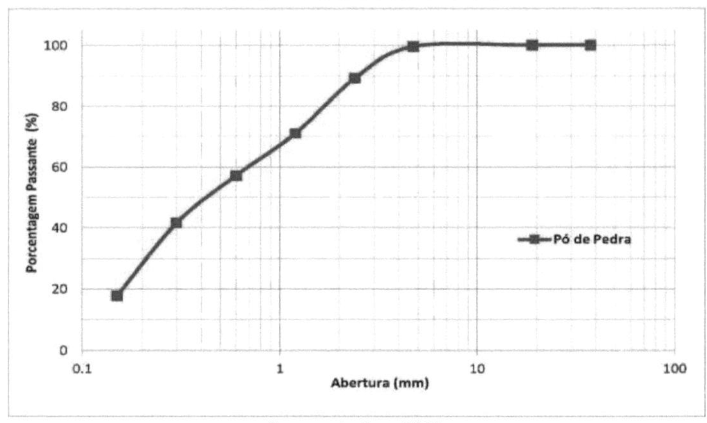

Source: Author, 2019.

Analysing Graph 1, it is possible to identify the representative diameters of the waste: D_{10} =

0.15 mm; D_{30} = 0.21 mm; D_{50} = 0.42 mm; D_{60} = 0.7 mm; D_{90} = 2.4 mm. The equivalent diameter (D_{10} = 0.15 mm) falls within the particle size range, which according to Bauer (2008) is equivalent to fine sand (0.15 mm - 0.6 mm). The highest percentage retained was in the 0.30 mm and 0.15 mm meshes, representing 48.00% of the total. This shows that the waste contains a large number of fine particles. It was also possible to obtain the Fineness Modulus (MF) from the accumulated retained percentage, which is equal to 2.24, so it can be inferred that the stone dust falls within the granulometric range, which according to ABNT NBR 7211/09 is equivalent to the optimum zone for use and at the same time classifies it as fine sand (2.20 < MF < 2.90).

The percentages retained and the particle size curve resulting from the sand particle size test in accordance with ABNT NM 248/03 are shown in Table 5 and Graph 2.

Table 5: Particle size composition of sand.

ABNT Sieve Opening (mm)	Retained percentage (%)	Cumulative Retained Percentage (%)	Passing percentage (%)
37,5	0,00	0,00	100,00
19	0,00	0,00	100,00
4,75	1,00	1,00	98,80
2,4	6,00	6,00	93,80
1,2	22,00	22,00	75,00
0,6	60,00	60,00	42,20
0,3	92,00	92,00	8,40
0,15	99,00	99,00	1,40
Fund	100,00	100,00	0,00

Source: Author, 2019.

Gráfico 2: Curva Granulométrica da Areia.

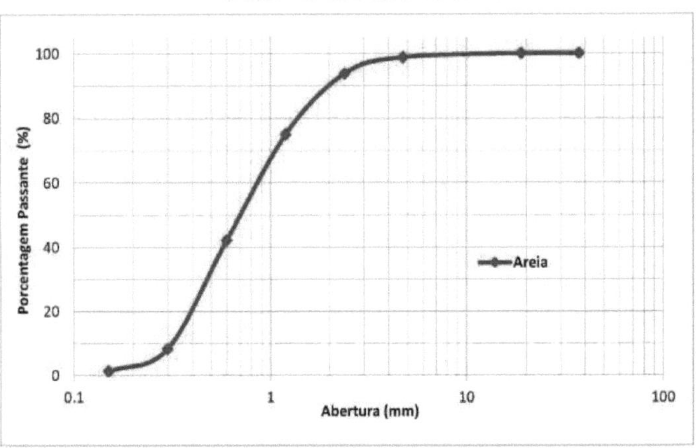

Fonte: Autor, 2019.

Analysing the granulometric composition and the granulometric curve of the sand (Table 4 and Graph 2), it can be seen that there is a considerable concentration of sand on the 1.2 mm and 0.6 mm mesh sieves, which correspond to 54.00% of the total, showing, according to Bauer (2008), that most of its grains have medium granulometry. Based on the data in Table 4, the Fineness Modulus (MF) of the sand was calculated, which corresponds to 2.79, placing it in the granulometric range, which according to ABNT NBR 7211/09 corresponds to the usable zone and at the same time classifies it as medium sand ($2.4 < MF < 3.2$). It is possible to see similarities between the particle size curves of sand and stone dust, as they have close quantities of grains in the range between 2.4 mm and 0.60 mm. However, there is a greater quantity of particles with a diameter of 1.2 mm in stone dust compared to sand, which may allow for the filling of voids with such dimensions, which are not fully supplied by natural aggregate.

Stone dust has a greater quantity of fines than sand (material passing the 0.15 mm sieve), which despite providing greater workability, requires a greater quantity of water in the mix, resulting in a reduction in concrete strength. The sand, in turn, contains a greater number of grains with dimensions equal to 0.60 mm and 0.30 mm when compared to the residue, which together with the stone dust particles provide a more effective filling of the voids.

Graph 3: Particle size curves - Sand x Stone Dust and limits.

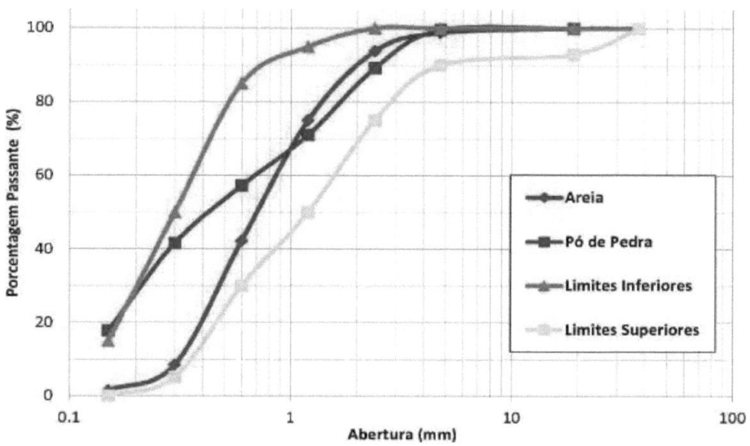

Source: Author, 2019.

It can be seen from Table 3 and Graph 1 that although the stone dust is in the granulometry range equivalent to fine sand, it still has a granulometry suitable for concrete, as does the sand, classified as medium sand, both of which are within the lower and upper limits established by ABNT NBR 7211/09, as shown in Graph 4.

Coarse Aggregate Particle Size

Table 6 shows the results of the granulometric composition of the gravel used in the research work.

Table 6: Grain size composition of gravel.

ABNT Sieve Opening (mm)	Retained percentage (%)	Cumulative Retained Percentage (%)	Passing percentage (%)
37,5	0,00	0,00	100,00
19	3,06	3,06	96,94
4,75	96,46	99,52	0,48
2,4	0,32	99,84	0,16
1,2	0,10	99,94	0,06
0,6	0,02	99,96	0,04
0,3	0,00	99,96	0,04
0,15	0,00	99,96	0,04
Fund	0,04	100,00	0,00

Source: Author, 2019.

Using the 19 mm sieve, a cumulative percentage of 3.06% was retained, which shows that the gravel used has a maximum diameter of 19 mm.

Graph 4: Grain size curve for gravel.

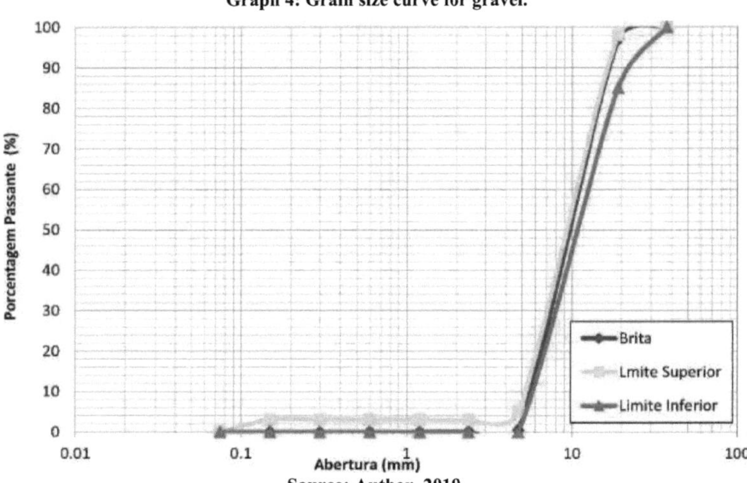

Source: Author, 2019.

Furthermore, according to the particle size curve drawn up with the accumulated retained percentages of the material (Graph 5), Brita 1 was found to be in favourable condition for concrete production, since the data found remained within the limits established by ABNT NBR 7211/09.

6.1.2 Unit mass, specific mass, fineness modulus and absorption

The results for the unit mass, specific mass, fineness and absorption of the materials are shown in Table 7.

Table 7: Physical characteristics of the materials.

Material	Unit Mass (g/cm)³	Specific Mass (g/cm3)	Fineness Modulus	Fineness (%)	Maximum diameter (mm)	Absorption (%)
Sand	1,53	2,60	2,79	-	2,4	-
Stone Dust	1,60	2,79	2,24	-	2,4	-
Brita	1,45	2,70	6,02	-	19,0	0,70
Binder	1,40	2,96	-	2,7	-	-

Source: Author, 2019.

The properties provided by the unit mass take into account the voids between the grains, based on the volume of the compacted aggregate. According to Cabral (2007), the unit mass is used to choose the mix of aggregates that will make the concrete more compact. The results presented in Table 7 show that the unit mass values of the sand and the material that replaces the stone dust are close. It can be seen that sand is lighter than stone dust, given its lower unit mass. Nóbrega (2007) states that because stone dust has a higher concentration of non-spherical particles, it will consequently have a characteristic lamellar structure.

The specific mass of stone dust is around 7.31 per cent higher than that of natural aggregate, which will probably result in a heavier end product. The higher unit mass of stone dust shows that, despite having a higher specific mass, the sand grains are better arranged, leaving fewer voids.

The specific mass of the aggregate is directly proportional to its fines content. Thus, the higher specific mass of stone dust compared to sand is indicative of the greater quantity of fine particles in the tailings, a fact proven by comparing the particle size curves of sand and stone dust (Graph 3).

The fineness modulus of the sand is higher than that of the stone powder, due to the greater number of larger particles in the sand (1.2 mm) compared to the residue. The latter, however, has a greater amount of fines (passing the 0.15 sieve) which makes the material more workable, but requires a greater amount of water in the mix. Stone dust has a lower fineness modulus compared to sand, due to the large amount of particles retained in the 0.30 mm and 0.15 mm meshes, which can help fill voids that natural aggregate grains are unable to fill.

The fineness of the cement obtained through the test was 2.7%, which is less than the maximum limit of 12% established by ABNT standard NBR 11579/13. According to Bauer (2003), this factor can influence the permeability of concrete. The finer the cement, the greater the amount of water consumed in the hydration process of the grains, reducing the excess water available for the formation of capillaries, thus improving the resistance, increasing the impermeability, workability and cohesion of concretes.

6.2 Mechanical tests

6.2.1 Simple compressive strength (SCR)

Table 8 shows the results of the simple compressive strength test of the specimens at 7, 14 and 28 days of submerged curing, carried out in accordance with ABNT NBR 5738/16.

Table 8: Simple compressive strength values at 7, 14 and 28 days of submerged curing.

Content of Replacement	RCS at 7 days (Mpa)	RCS at 14 days (Mpa)	RCS at 28 days (Mpa)
0%	17,11 ± 0,27	18,52 ± 0,67	20,19 ± 0,28
25%	19,86 ± 1,24	21,95 ± 0,23	24,10 ± 1,02
50%	20,35 ± 0,45	23,12 ± 0,05	23,44 ± 1,51
75%	20,65 ± 1,08	24,24 ± 0,91	26,77 ± 0,70
100%	22,12 ± 1,54	24,72 ± 1,14	28,12 ± 0,83

Source: Author, 2019.

In order to show a better understanding and visualisation of the values in Table 8, Graph 6 shows the values grouped together, where an increase in resistance over time is noticeable.

Graph 5: RCS values at 7, 14 and 28 days of curing.

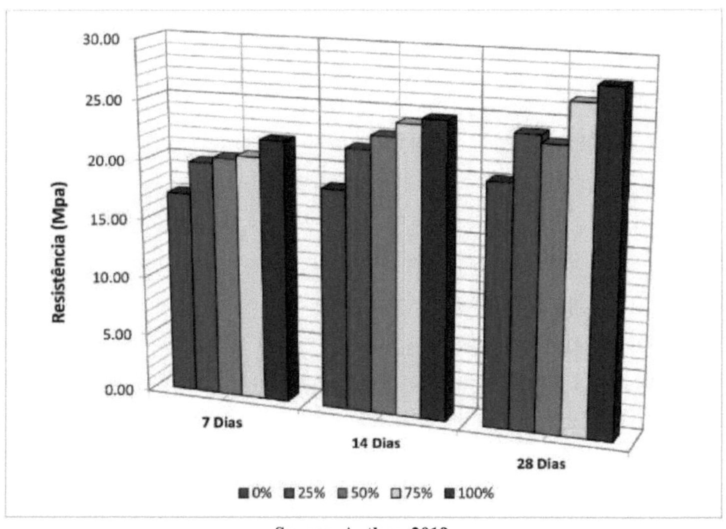

Source: Author, 2019.

When analysing the results for simple compressive strength, it can be seen that when comparing the reference mix (0% waste) with the mix containing 100% stone dust, the increase in strength is significant. At 7 days, this increase was of the order of 29.3%, at 14 days the concrete had increased in strength by 33.5% and at 28 days it was 39.3% stronger than the comparative composition.

Simple compressive strength (SCR) values were also found to be similar to the work carried out by Menossi (2004), who obtained satisfactory results in terms of the increase in concrete compressive strength with the use of stone dust. However, the author's tests showed an increase in simple compressive strength at 28 days of curing of around 66%, a much higher value than that obtained in this study.

Graph 7 below shows the rebates obtained for each composition, in accordance with ABNT NBR NM 67/98, with the w/c ratio kept constant and equal to 0.70.

Graph 6: Rebates and their respective compositions.

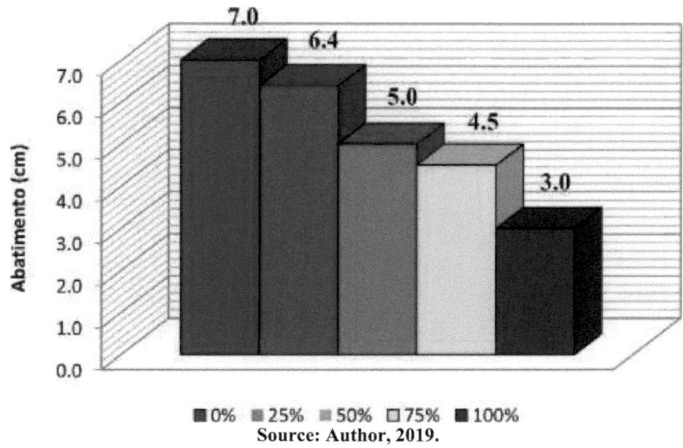

Source: Author, 2019.

When analysing the simple compressive strength results shown in Table 8 and Graph 6, it can be seen that at 28 days all the compositions reached or exceeded the minimum strength limit for concrete for structural purposes, which is 20 Mpa according to NBR 6118/2014. Likewise, we can infer that the composition with 100% stone dust had the highest strength at all curing ages; on the other hand, it had a very low slump. This high resistance is not only linked to a low slump, but also to the presence of fines, which provide greater compactness in the concrete, reducing its porosity, making it more resistant.

Even though the slump decreased as the substitution of sand for stone dust increased, higher strength values were found compared to the study carried out by Bastos (2003). The author also obtained satisfactory results in terms of simple compressive strength (SCR), but with a low slump. The same author studied substitutions of natural sand for stone dust of between 15 and 70 per cent by mass in the production of conventional concrete. However, for both studies, the higher the stone dust content used, the lower the slump and the higher the simple compressive strength (SCR), and the best performance was observed for the highest levels of substitution of natural sand for stone dust in both studies.

Concrete with added stone dust showed a gain in strength in all compositions and curing ages, as shown in Graph 8 below.

Graph 7: Comparison between the Simple Compression Strengths obtained.

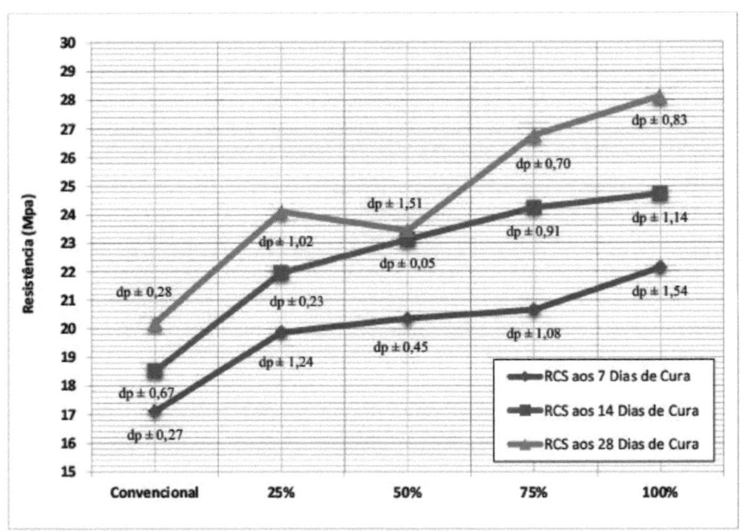

Source: Author, 2019.

sd: standard deviation

Analysing Graph 8, it can be seen that the compositions with 0%, 25% and 50% substitution of sand by stone dust, respectively, had a proportional increase in strength from 7 to 28 days, with the exception of the composition with 50% waste which showed a slight decrease in strength at 28 days of curing, reaching a maximum variation of around 13.6%, while the compositions with 75% and 100% substitution, respectively, had a greater gain in strength at 28 days. This gain in strength of the compositions with higher levels of stone dust can be explained by the characteristics of these aggregates, as stated by Leite (2001); Neville (1997), that waste properties such as texture and granulometry positively influence the increase in strength. In concrete with the replacement of natural fine aggregate with waste, all these aspects are significant in increasing strength.

In order to better analyse the use of stone dust in concrete, an open-air curing test was carried out, where specimens from all the compositions studied, 0%, 25%, 50%, 75% and 100% replacement of sand by stone dust, were exposed to the sun and rain without any protection and moulded together with those previously analysed.

Table 9 shows the RCS obtained by the specimens with submerged curing and open-air curing at 28 days.

Table 9: Comparison between RCS.

Substitution content	RCS (Mpa)	
	Submerged Curing	Outdoor Healing
0%	21,09 ± 0,28	15,57 ± 1,31

25%	24,10 ± 1,02	16,85 ± 1,26
50%	23,44 ± 1,51	17,22 ± 1,35
75%	26,77 ± 0,70	19,77 ± 0,97
100%	28,12 ± 0,83	21,90 ± 1,42

Source: Author, 2019.

Graph 8: Comparison between Submerged Curing and Open Air Curing.

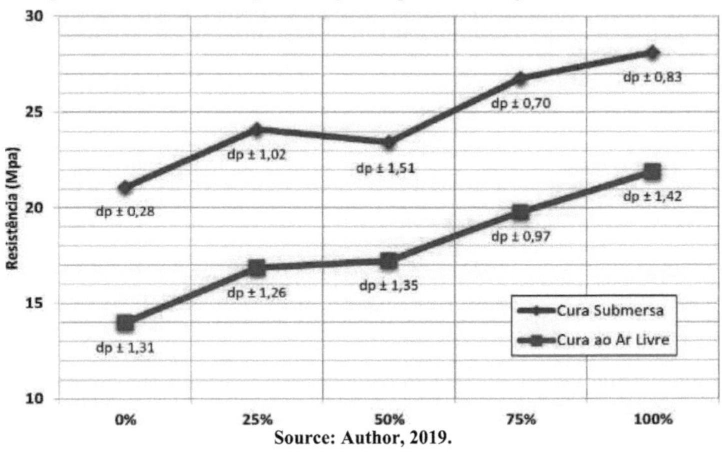

Source: Author, 2019.

sd: standard deviation

Analysing Table 9 and Graph 9, it can be seen that at 28 days the composition with 100% stone dust, even when cured in the open air, i.e. without any kind of protection from the weather, such as sun or rain, which can lead to deficiencies in the concrete, had a higher strength than the reference composition, cured underwater. According to Menossi (2004), in his study of a concrete that used stone dust in its composition to replace natural sand, he observed an increase in strength that compensated for the unfavourable curing conditions that the concrete was subjected to in his study. The author's tests showed a 26.00% increase in compressive strength in concrete cured under unfavourable conditions compared to normal concrete with submerged curing, a higher increase than that obtained in the present study. Even so, even though the strength values were not higher than in Menossi's study, there was a similar behaviour to that presented by the author, in terms of a gain in strength proportional to the residue content in the mix.

Menossi (2004) also points out that this is of great importance, because in Brazil, most small construction sites do not have any type of curing control for the concrete, with the master builder or bricklayer, at most, responsible for wetting the surface of these concretes a few times a day. Therefore, concrete with stone dust has a further advantage, which is that it achieves satisfactory strengths even without any specific curing.

6.2.2 Water absorption

Table 10 shows the average results of the water absorption test of the concrete studied in its different proportions of replacement of the fine aggregate with stone dust, at 7, 14 and 28 days of submerged curing, carried out in accordance with ABNT NBR 9778/09.

Table 10: Water absorption test results.

Substitution content	Absorption at 7 Curing Days (%)	Absorption at 14 Curing Days (%)	Absorption at 28 Curing Days (%)
0%	5,91 ± 0,30	4,73 ± 0,40	3,06 ± 0,53
25%	6,14 ± 0,47	4,95 ± 0,49	3,19 ± 0,36
50%	6,42 ± 0,83	5,11 ± 0,85	3,31 ± 0,43
75%	6,89 ± 0,65	5,30 ± 1,20	3,46 ± 1,35
100%	7,21 ± 0,31	5,58 ± 0,69	3,52 ± 0,73

Source: Author, 2019.

The absorption results shown in Table 10 show that concrete with stone dust generally had higher absorption values than the reference concrete for all compositions at the three curing ages. This is because the residue has a greater quantity of fines that absorb more water than sand, a fact that can be seen by analysing Graph 5, which relates the particle size curves of sand and stone dust.

In addition, the geometry of the grains influences the amount of water it will absorb. According to Londero *et al.* (2017), a large number of non-spherical particles provide a greater volume of voids, which leads to greater absorption. Thus, due to its lamellar shape, stone dust ends up producing a more porous concrete, i.e. a concrete with greater water absorption. The same fact was verified by Levy (2001) in his analysis of the behaviour of concrete produced with fine aggregate from the rock crushing process. The author found that as the natural aggregate is replaced by waste, the absorption value increases and can reach more than 60 per cent in relation to the reference concrete, depending on the proportion of replacement and the waste chosen.

Graph 9: Comparison of water absorption values.

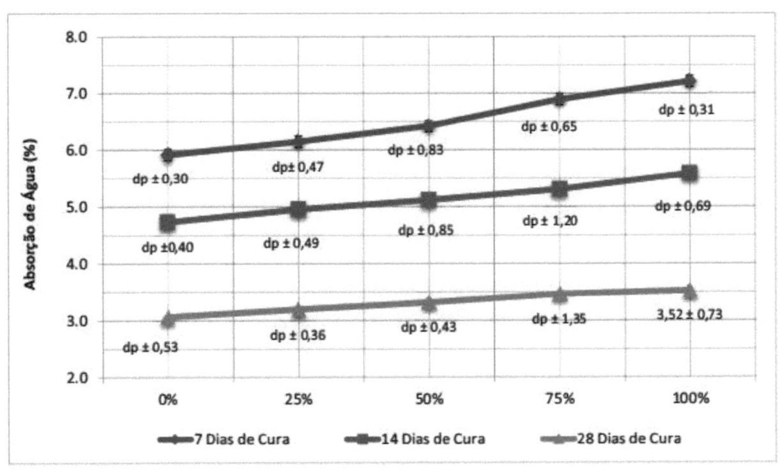

Source: Author, 2019.

sd: standard deviation

Graph 10 shows a certain linearity in the results obtained from the water absorption test at all the curing ages studied, regardless of the proportion of substitution, but none of the substitutions studied achieved absorption lower than or similar to the reference composition, made up of natural aggregate only. This variation in absorption was greatest in concrete with a 100% replacement ratio at 7 days of curing, reaching a 22.00% increase in absorption compared to the reference concrete. From then on, as the curing time increased, there was a decrease in this variation in water absorption. At 14 and 28 days of curing, the maximum absorption results achieved were 17.97% and 15.00%, respectively. In short, there is a linear increase in water absorption proportional to the percentage of waste substituted for sand, i.e. as the proportion of waste substituted increases in the mix, absorption increases proportionally. This is proven by the fact that the concrete with the highest stone dust content (100% substitution) obtained the highest absorption value.

Table 10 also shows that, at 7 days of curing, all the concretes had water absorption values higher than 6.2%, with the exception of the composition with 25% replacement and the reference concrete. According to ABNT NBR 9778/09, concretes with water absorption values higher than 6.2% are classified as deficient and may present problems in terms of their ability to withstand mechanical stress. At 14 days of curing, all the proportions studied had water absorption values in the range of 6.2% ≤ to ≤ 4.2%, which, according to ABNT NBR 9778/09, classifies them as conventional concrete. However, at 28 days of curing, all the proportions obtained water absorption results of less than 4.2%, which allows these concretes to be placed in the durable (high performance) class, according to the limits established by

ABNT NBR 9778/09.

Therefore, greater absorption is closely linked to lower strength, because a greater amount of water absorbed is accompanied by a higher void content, which reduces the compactness of the mix and, consequently, the strength of the concrete. This fact was found by Tenório (2007) who found lower simple compressive strength values for concrete with waste added to its composition than the reference concrete and attributed this behaviour to the low compactness and consequent porosity of the mixture caused by the artificial aggregate.

As shown in the previous stage of this study, the results obtained showed a different trend, since the concretes with the highest compressive strength were also those that had the highest water absorption. This particular behaviour was also verified by Rodrigues (2001), who found an increase in the simple compressive strength of concrete with waste compared to the reference concrete, as high water/cement ratios were used, and attributes this behaviour to the high absorption power of waste when acting as aggregates, since a portion of the water added during dosing is absorbed by the waste during mixing, as the artificial aggregate was not pre-wetted. As a result, the amount of water absorbed by the waste can promote "internal curing" during the hardening of the paste, providing better results in terms of resistance to simple compression. According to the author, increasing the w/c ratio tends to reduce the difference between the strength of conventional concrete and concrete with waste.

Thus, according to the results obtained, the vast majority of the concretes produced fall within acceptable usage ranges in terms of water absorption.

CONCLUSIONS

From the results obtained in the stages of this research work, it can be concluded that:

- The stone dust residue and the sand had a close number of grains in the 2.4 mm to 0.60 mm range. Natural aggregate, on the other hand, contains a greater number of grains with dimensions equal to 0.60 mm and 0.30 mm when compared to waste. The tailings have a higher percentage retained in the 0.30 mm and 0.15 mm meshes.
- The unit and specific masses of the aggregates\ studied showed that both are similar in these properties.
- For the stone dust studied, the characterisation tests carried out showed that this material has all the physical characteristics necessary for its use as a fine aggregate in concrete. Its particle size curve, according to ABNT NBR 7211/09, classifies it as fine sand.
- Concrete with stone dust showed an increase in compressive strength compared to concrete made with natural sand. At 28 days, this increase totalled 40.5%.
- Concrete with 100% stone dust showed excellent behaviour at all the curing ages studied. At 7 days, concrete with 100% stone dust showed an increase of approximately 29.3% compared to normal concrete, and these values increased to 33.5% at 14 days and 40.5% at 28 days.
- In terms of water absorption, all the compositions with the addition of stone dust showed an increase in absorption compared to the reference composition, made up only of natural fine aggregate. The composition with 100 per cent waste had higher absorption and higher resistance to simple compression (RCS).

In view of the above, the steps proposed for this work demonstrate the technical feasibility and advantages of replacing natural aggregate with stone dust in usual concretes, since the values obtained in the mechanical tests were satisfactory for all the compositions evaluated, at all the curing ages studied, as well as contributing to the reduction of environmental impacts in the concrete production process.

SUGGESTIONS FOR FUTURE WORK

Bearing in mind that this work was limited to concretes with usual strengths and that its objective was to study stone dust as a fine aggregate for concrete, there are several studies that can be carried out based on the conclusions presented here:

- Study on the use of stone dust in laying mortars;
- Study of the influence of increasing the water cement ratio for concretes containing stone dust;
- Study on the use of stone dust in high-performance concrete (CAD) and high-strength concrete (CAR);
- Study on the use of stone dust in self-compacting concrete;
- Study on the use of stone dust in conjunction with an additive for

concrete production, carrying out a detailed study of the cost/benefit ratio resulting from the use of the additive.

BIBLIOGRAPHICAL REFERENCES

ABESC - Associacao brasileira das empresas de servicos de CONCRETAGEM DO BRASIL. **Manual** do **concreto dosados em central.** Available at <http://www.abesc.org.br/pdf/manual.pdf> Accessed on 27 November 2018.

ABNT - Brazilian Association of Technical Standards, **NBR 11579: Portland cement - Determination of fineness by means of the 75 µm sieve (n° 200).** Rio de Janeiro, 2013.

ABNT - Brazilian Association of Technical Standards, **NBR 12655: Portland Cement Concrete - Preparation, Control, Receipt and Acceptance - Procedure.** Rio de Janeiro, 2015.

ABNT - Brazilian Association of Technical Standards, **NBR 16605: Portland cement and other powdered materials - Determination of specific mass.** Rio de Janeiro, 2017.

ABNT - Brazilian Association of Technical Standards, **NBR 5738: Concrete - Procedure for Moulding and Curing and Test Specimens.** Rio de Janeiro, 2016.

ABNT - Brazilian Association of Technical Standards, **NBR 5739: Concrete - Compression Test of Cylindrical Specimens.** Rio de Janeiro, 2018.

ABNT - Brazilian Association of Technical Standards, **NBR 7211: Aggregates for concrete - Specification.** Rio de Janeiro, 2009.

ABNT - Brazilian Association of Technical Standards, **NBR 9778: Hardened mortar and concrete - Determination of water absorption by immersion - Void index and specific mass.** Rio de Janeiro, 2009.

ABNT - Brazilian Association of Technical Standards, **NBR NM 248: Aggregates - Determination of granulometric composition.** Rio de Janeiro, 2003.

ABNT - Brazilian Association of Technical Standards, **NBR NM 52: Aggregate - Determination of Specific Mass and Apparent Specific Mass.** Rio de Janeiro, 2009.

ABNT - Brazilian Association of Technical Standards, **NBR NM 45: Aggregates - Determination of unit mass and volume of voids.** Rio de Janeiro, 2006.

AGOPYAN, V. **Construction consumes up to 75 per cent of the planet's raw materials. In: GloboCiência,2013.**Available at:<http://redeglobo.globo.com/globociencia/noticia/2013/07 /construção-civil-consome-ate-75-da-materia-prima-do-planeta.html>. Accessed on: 02 March 2019.

AGOPYAN, Vahan; JOHN, Vanderley M**. O desafio da sustentabilidade na construção**. São Paulo: Blucher, 2011.

ANDRIOLO, F. R. **Concrete Constructions**. São Paulo: Pini, 1984. 738 p.

ANDRIOLO, F. R. **Uses and abuses of stone dust in different types of concrete**. In: Seminar: The use of fine crushed stone. II SUFFIB, São Paulo, 2005. **Proceedings,** São

Paulo, EPUSP, 2005.

Annual Book of ASTM Standards. **American Society for Testing and Materials**. ASTM C1260/94-Standard Test Method for potential alkali reactivity of aggregates (mortar-bar method). Philadelphia, 1997.

BASTOS, S. R. B. **Use of artificial basalt sand as a partial substitute for fine sand in the production of conventional concrete**. In: 45th Brazilian Concrete Congress - IBRACON. Proceedings: Brazilian Concrete Institute, Vitória, August 2003.

BATTAGIN, A.F. A **brief history of Portland cement**. Available at <http:// www.abcp.org.br/basico_sobre_cimento/historia.shtml>. Accessed on: 02 March 2019.

BAUER, L. A. F. **Materiais de Construção**. Rio de Janeiro, LTC - Livros Técnicos e Científicos, 5th Ed, v. 1, 2003. 435p.

BAUER, L. A. F. **Materiais de Construção.** 5. ed. Rio de Janeiro: Ltc, 2008. 538 p.

CABRAL, A. E. B. **Modelling of mechanical properties and durability of concrete produced with recycled aggregates, considering the variability of the composition of the CDW**. São Carlos: USP. 2007. Master's dissertation. São Carlos School of Engineering.

CARLOS, E. M. **Effect of the addition of scheelite residue on the technical-mechanical and rheological behaviour of mortars for ceramic engobes**. 2018. 122f. Thesis (PhD in Mechanical Engineering) - Technology Centre, Federal University of Rio Grande do Norte, Natal, 2018.

COSTA, MJ. **Evaluation of the use of artificial sand in Portland cement concrete: Applicability of a dosage method**. Civil Engineering Course Conclusion Work - UNIJUÍ, 2005.

DIAS, S. L. **Incorporation of kaolin waste in mortars for laying and coating for use in civil construction - evaluation of pozzolanic activity.** 2010. 36 f. Dissertation (Master's Degree) - Postgraduate Course in Materials Science and Engineering, Centre for Science and Technology, Federal University of Campina Grande, Campina Grande, 2010.

GIAMMUSSO, S., **Concrete Manual**. São Paulo: PINI, 1992, p. 23-24.

HELENE, Paulo; TERZIAN, Paulo. **Manual de dosagem e controle do concreto**. Ed. PINI, 1ª edition. São Paulo, 1995.

IBGE. **Annual survey of the construction industry.** Rio de Janeiro, 2007. Available at: <http://www.ibge.gov.br/home/estatistica/economia/industria/paic/2007/default.shtm>. Accessed on: 28 February 2019.

IBGE. **Annual survey of the construction industry.** Rio de Janeiro, 2016. Available at: <http://www.ibge.gov.br/home/estatistica/economia/industria/paic/2016/default.shtm>. Accessed on: 28 February 2019.

JACQUES, J. R.. **Study of the technical feasibility of using recycled concrete as aggregate in Portland cement concrete.** 2013. 32 f. TCC (Graduation) - Civil Engineering Course, Department of Exact Sciences and Engineering, Universidade Regional do Noroeste do Estado do Rio Grande do Sul, Ijuí, 2013.

KAIFFER, L. F. **The evolution of reinforced concrete**. São Paulo, 1998.

LEITE, M. B. **Evaluation of the mechanical properties of concrete produced with recycled aggregates from construction and demolition waste**. 2001. 266 p. Thesis (Doctorate in Engineering), School of Engineering, Postgraduate Programme in Civil Engineering, Federal University of Rio Grande do Sul, Porto Alegre, 2001.

LEVY, S. M. **Contribuição ao estudo da durabilidade de concretos, produzidos com resíduos de concreto e argamassa**. São Paulo: USP, 2001. Thesis (Doctorate) - Polytechnic School of the University of São Paulo.

LONDERO, C. **Determination of the packing density of composite granular systems from IPT normal sand: comparison between particle size distribution optimisation models and random compositions. Cerâmica**, [s.l.], v. 63, n. 365, p.2233, mar. 2017.

MARTIN, J. F. M. Additives for Concrete. In: ISAIA, Geraldo Cechella. **Concrete: Teaching, Research and Realisations**. São Paulo: Editora IBRACON, 2005, V1. Chap. 13, p. 381-406.

MEHTA, P. K., MONTEIRO, P. J. M. **Concreto - Estrutura, Propriedades e Materiais**. Ed. PINI, 1st edition. São Paulo, 1994.

MEHTA, P. K.; MONTEIRO, PAULO J. M. **Concreto - Estrutura, Propriedades e Materiais**. 3ª . Ed. São Paulo: Ibracon, 2008.

MEHTA, P.K.; MONTEIRO, P.J.M. **Concreto: Estrutura, Propriedades e Materiais**. São Paulo: PINI, 1997.

MENOSSI, R. T. **Utilisation of Basalt Stone Powder as a Substitute for Natural Sand in Concrete.** Dissertation (Master's in Civil Engineering) - Faculty of Engineering, Ilha Solteira Campus, Paulista State University, Ilha Solteira, 2004.

NETO, C. S. Aggregates for Concrete. In: ISAIA, Geraldo Cechella. **Concrete: Teaching, Research and Achievements**. São Paulo: IBRACON Publishing House, 2005. V1, Chap. 11, p. 323-326.

NEVILLE, A.M. **Properties of Concrete**. Editora Pini, São Paulo, 1982.

NEVES, C. A. R.; SILVA, L. R. **Universo da Mineração Brasileira**, 2007.

NÓBREGA, A. F. **Potential use of kaolin waste from Paraíba for the development of multi-purpose mortars.** 2007. 59 f. Dissertation (Master's Degree) - Postgraduate Course in Urban Engineering, Technology Centre, Federal University of Paraíba, João Pessoa, 2007.

RODRIGUES, C. R. S. **Evaluation of the mechanical properties of concrete produced with recycled construction waste aggregate**. (Master's dissertation). 2011. POLI. Recife-PE.

SILVA, S. H. **Analysis of reinforced concrete structures subject to corrosion of reinforcement by chlorides using the finite element method**. Porto Alegre, 2004. 223 p. Dissertation (Master's in Engineering), Postgraduate Programme in Civil Engineering, UFRS.

SOUZA, P. A. **Study of the plastic, mechanical, microstructural and thermal behaviour of concrete produced with porcelain tile waste**. Natal, 2007. PhD Thesis, Postgraduate Programme in Materials Science and Engineering, UFRN.

TENÓRIO, J. J. L. **Evaluation of the properties of concrete produced with recycled aggregates from construction and demolition waste for structural applications** (Master's thesis). 2007. UFAL. Maceió-AL.

THEISEN, J. H. **Production and destination of urban solid waste in the city of Itapiranga- Brazil. Ijuí,** 2012. 26 f. TCC (Graduation) - Geography Course, Universidade Regional do Noroeste do Estado do Rio Grande do Sul, Ijuí, 2012.

TIECHER, F.; PANDOLFO, L.M. **Concreto: Comparison of experimental dosing methods using artificial sand**. In: 46th Brazilian Concrete Congress - IBRACON. **Proceedings**: Brazilian Concrete Institute, Florianópolis, 2004.

VIEIRO, E. H. **Application of basalt rock crushing sand in the manufacture of Portland cement concrete**. Dissertation presented to the Postgraduate Programme in Materials at the University of Caxias do Sul, 2010.

Printed by Books on Demand GmbH, Norderstedt / Germany